MW01593244

JUL   2022

Sea Horses

by Kate Moening

BELLWETHER MEDIA • MINNEAPOLIS, MN

BLASTOFF!
2
READERS

Blastoff! Readers are carefully developed by literacy experts to build reading stamina and move students toward fluency by combining standards-based content with developmentally appropriate text.

Level 1 provides the most support through repetition of high-frequency words, light text, predictable sentence patterns, and strong visual support.

Level 2 offers early readers a bit more challenge through varied sentences, increased text load, and text-supportive special features.

Level 3 advances early-fluent readers toward fluency through increased text load, less reliance on photos, advancing concepts, longer sentences, and more complex special features.

★ **Blastoff! Universe**

Reading Level

Grade **K**

Grades **1–3**

Grade **4**

This edition first published in 2022 by Bellwether Media, Inc.

No part of this publication may be reproduced in whole or in part without written permission of the publisher. For information regarding permission, write to Bellwether Media, Inc., Attention: Permissions Department, 6012 Blue Circle Drive, Minnetonka, MN 55343.

Library of Congress Cataloging-in-Publication Data

Names: Moening, Kate, author.
Title: Sea horses / Kate Moening.
Description: Minneapolis, MN : Bellwether Media, [2022] | Series: Blastoff! readers : Animals of the coral reef | Includes bibliographical references and index. | Audience: Ages 5-8 | Audience: Grades 2-3 | Summary: "Relevant images match informative text in this introduction to sea horses. Intended for students in kindergarten through third grade"--Provided by publisher.
Identifiers: LCCN 2021000530 (print) | LCCN 2021000531 (ebook) | ISBN 9781644875056 (library binding) | ISBN 9781648344138 (ebook)
Subjects: LCSH: Sea horses--Juvenile literature.
Classification: LCC QL638.S9 M64 2022 (print) | LCC QL638.S9 (ebook) | DDC 597/.6798--dc23
LC record available at https://lccn.loc.gov/2021000530
LC ebook record available at https://lccn.loc.gov/2021000531

Editor: Elizabeth Neuenfeldt   Designer: Laura Sowers

Printed in the United States of America, North Mankato, MN.

Table of Contents

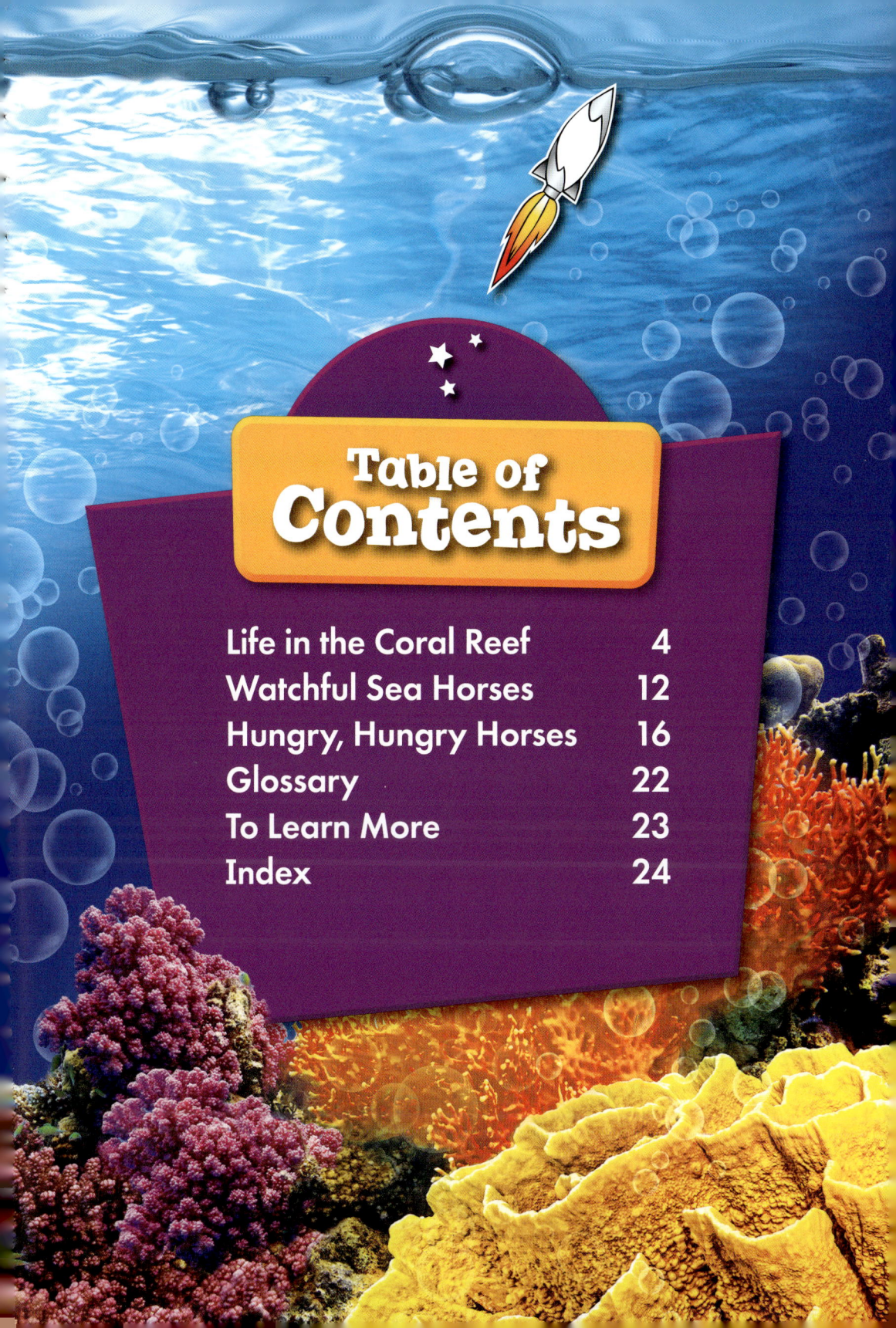

Life in the Coral Reef

tiger tail sea horse

Sea horses are named after their shape. They have horse-like heads!

These fish live in the coral reef **biome**.

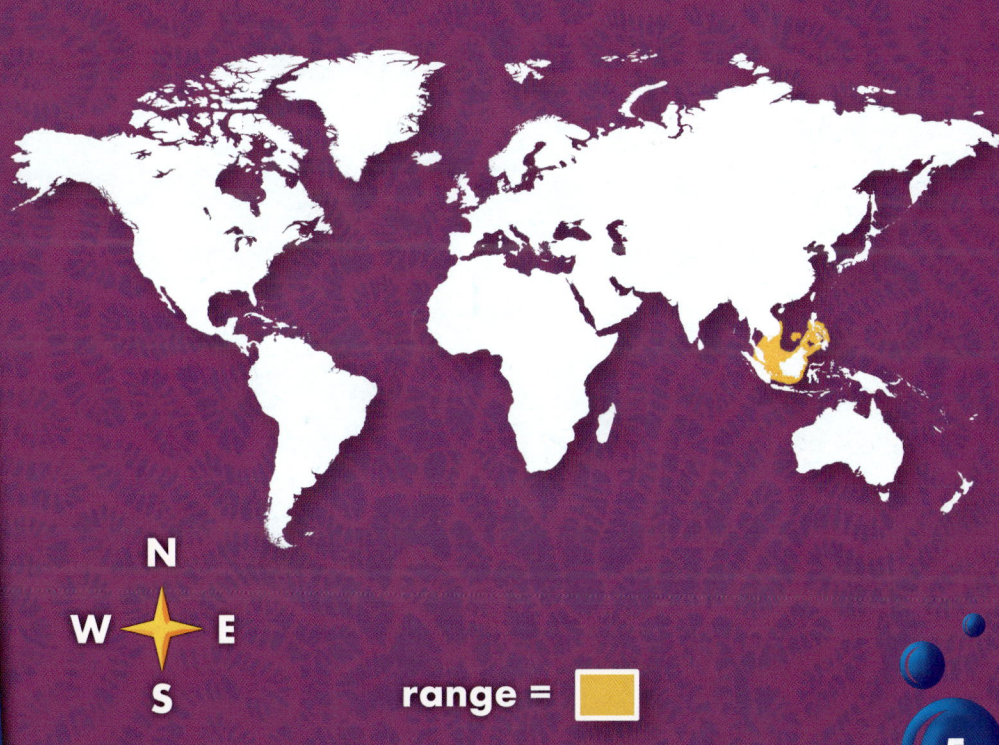

N
W ✦ E
S

range =

5

Sea horses are built to survive coral reefs. They have excellent **camouflage**!

Caribbean dwarf sea horse

**thorny
sea horse**

Some blend in with sea grass.
Others are colorful like **corals**.

Sea horses have hard bodies!
They are covered in bony plates.

Special Adaptations

bony plates

long snout

strong tail

The plates help protect sea horses from **predators**.

Coral reefs can have strong **currents**.
Sea horses use their tails to stay in place.

Their tails hold on tight to sea grasses and corals!

Tiger Tail Sea Horse Stats

Least Concern	Near Threatened	Vulnerable	Endangered	Critically Endangered	Extinct in the Wild	Extinct

conservation status: vulnerable

life span: about 2.5 years

Watchful Sea Horses

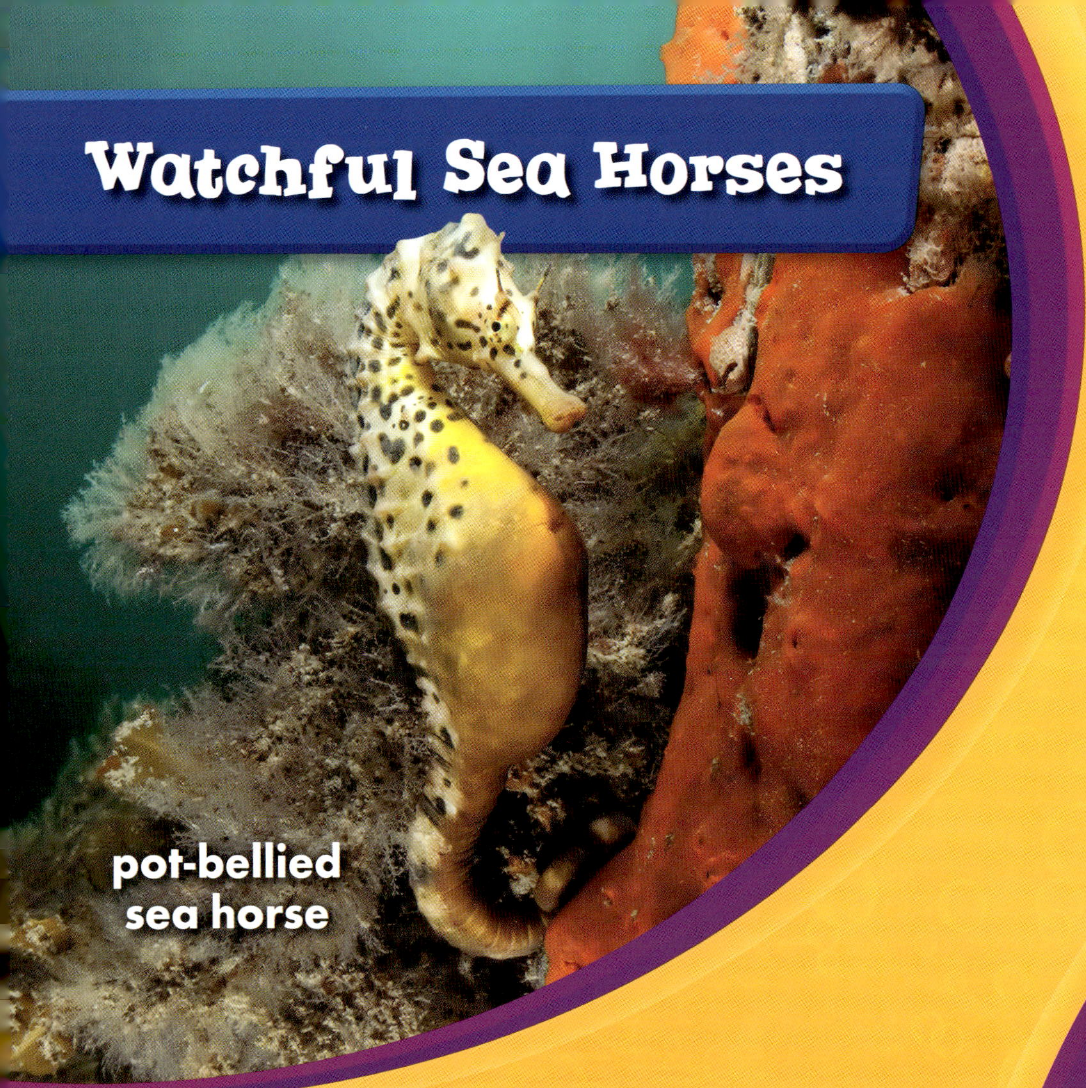

pot-bellied
sea horse

Sea horses have great eyesight. They can easily spot predators and food.

These fish can move their eyes in **opposite** directions!

yellow Pacific
sea horse

13

Sea horses are slow swimmers.

If they spot a predator, sea horses change colors to blend in! This helps sea horses stay safe.

Hungry, Hungry Horses

Sea horses do not have stomachs! These fish quickly **digest** food.

Sea horses must eat all day to stay healthy.

Sea horses hide and stay still to surprise **prey**. This helps sea horses save **energy**!

Sea Horse Diet

ghost shrimp

brine shrimp

zooplankton

They mostly eat shrimp and **plankton**.

Sea horses suck up food with their long **snouts**. Their snouts also help them find food in tiny spaces.

snout

These fish have **adapted** well to their coral reef home!

Glossary

adapted—changed over a long period of time

biome—a large area with certain plants, animals, and weather

camouflage—coloring or markings that make animals look like their surroundings

corals—the living ocean animals that build coral reefs

currents—flowing water that is always moving in the same direction

digest—to change food into a form that the body can use

energy—the power to move and do things

opposite—completely different

plankton—very small animals and plants that live in oceans; zooplankton are tiny animals and phytoplankton are tiny plants.

predators—animals that hunt other animals for food

prey—animals that are hunted by other animals for food

snouts—long noses

To Learn More

AT THE LIBRARY

Bozzo, Linda. *How Sea Horses Grow Up*. New York, N.Y.: Enslow Publishing, 2020.

Neuenfeldt, Elizabeth. *Sea Horse*. Minneapolis, Minn.: Bellwether Media, 2021.

Terp, Gail. *Seahorses*. Mankato, Minn.: Black Rabbit Books, 2021.

ON THE WEB

FACTSURFER

Factsurfer.com gives you a safe, fun way to find more information.

1. Go to www.factsurfer.com.

2. Enter "sea horses" into the search box and click 🔍.

3. Select your book cover to see a list of related content.

Index